young engineer

IN THE FACTORY

Malcolm Dixon

A Young Engineer Book

The Bookwright Press
New York · 1983

Young Engineer Series

In the Air

On the Road

In the Factory

In Communication

First published in the United States in 1983 by
The Bookwright Press, 387 Park Avenue South,
New York, NY 10016
First published in 1983 by
Wayland Publishers Limited, England
ISBN: 0-531-04701-6
Library of Congress Catalog Card Number 83-71638

Illustrated by Roger Norris
Printed in Italy by G. Canale & C.S.p.A.

Contents

Chapter 1 Looking at factories

Types of factories **4**
Inside a factory **5**

Chapter 2 Power in the factory

Water power **6**
Wind power **7**
Steam power **8**
Electric power **9**
Make an electric motor **10**
Compressed-air power **12**

Chapter 3 Transferring power

Using levers **13**
Using gears **14**
Using belt drives **16**
Using chain drives **18**

Chapter 4 Factory machines

A drop hammer **19**
Drilling **20**
Lathes **21**
Robots **22**

Chapter 5 Moving goods

Cranes **24**
Fork-lifts **26**
Conveyors **28**

Chapter 6 Factory problems

Friction **30**
Keeping cool **31**
Safety **32**

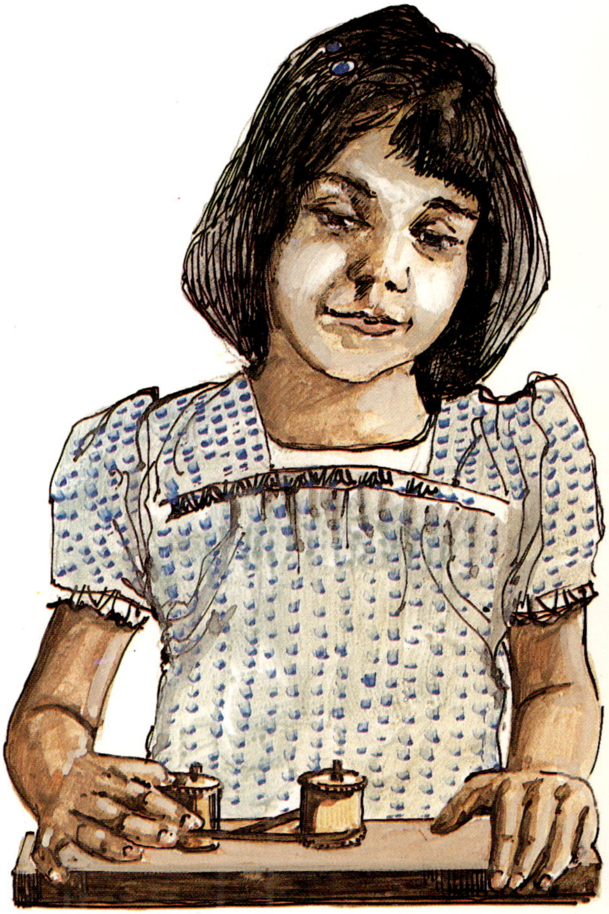

Chapter 1 Looking at factories

Types of factories

In our world, most of the things we use are made in factories. In the past skilled craftsmen used their hands to make things. Today people work in factories and use machines to make a product. This product may be a television set, a car, a table or a plastic bottle. Make a list of the products in your home that may have been made in a factory.

Look for factories near your home and school. Try to find out what is made in these factories. Are they making similar products or many different products?

Inside a factory

The skilled craftsworker often makes a complete product alone. In a modern factory one person may use a machine to make only a small part of a product. Other people using different machines will make the other parts. Some people may repeat the same actions all day. This system means that a product can be made cheaply and quickly.

Engineers help to build factories. They invent and build machines and help to look after them. They make sure that the machines have the power needed to work properly. See if your teacher or your parents can arrange a visit to a factory near you.

Chapter 2

Power in the factory

Water power

The machines in a factory need power to make them work. For hundreds of years flowing water in streams and rivers provided this power. A factory was built along the bank of a river. The water of the river turned a large water wheel. This moved the machinery inside the factory.

In some countries water is used to make electricity. Water is collected behind a dam and is then allowed to flow through special water wheels, called turbines. These turbines turn very quickly and are used to make electricity which can be used in factories. Engineers hope, also, to use the waves and tides of the sea to make power for factories.

Can you make some simple water wheels? Use corks, knitting needles, strips of plastic, wood and yogurt containers. How will you test your water wheels? Which type turns the fastest?

Wind power

At sea, wind power is used to move sailing boats. In areas without fast-flowing rivers, the wind has been used to provide factory power.

The early windmills were used to grind corn or pump water. Moving air turned the sails of the windmill and these turned machinery inside the mill.

Today, engineers are building new types of windmills to make electric power.

Make a model windmill. Use a plastic bottle, filled with sand, for the mill. Make the sails from a square piece of cardboard.
Test your windmill in a breeze. Do the sails turn?
Use other materials for the sails. Which works the best?

Steam power

The invention of the steam engine brought great changes to factories. In this machine, water is boiled and turns to steam. The steam is used to move pistons. The pistons move other engine parts and this power is used to work other machines. Modern power stations use steam to drive turbines, which are used to make electricity.

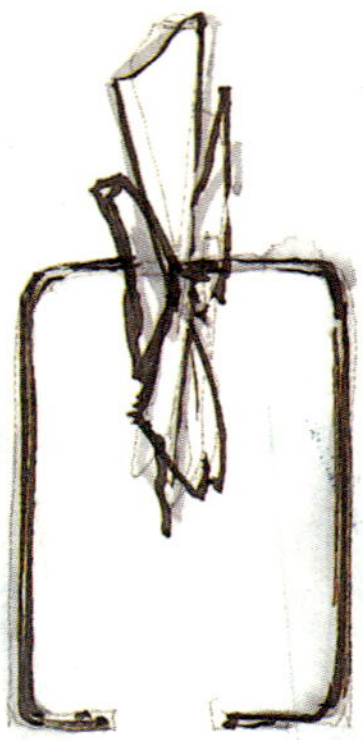

Find a metal cigar tube. Position the tube across two corks which have been glued to a wooden base. Straighten out some paperclips and use them to hold the tube in place. Pour a little water into the tube and fit a cork. Make a small hole in the cork. Make a small rotor from aluminum foil. Use a paperclip to fix this to the cork. **Ask an adult** to heat the water in the tube with a small candle. **Remember, steam can burn.** What happens to the rotor when the water is boiling?

Electric power

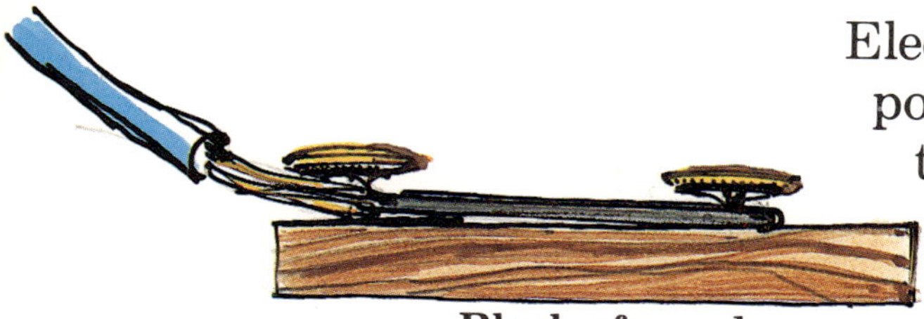

Electricity is a very convenient source of power. It is carried from power stations to factories by a network of cables. In the factory, electricity is used for lighting and heating. This means that people can work in pleasant and comfortable conditions. It also provides the power to run the machines in the factory.

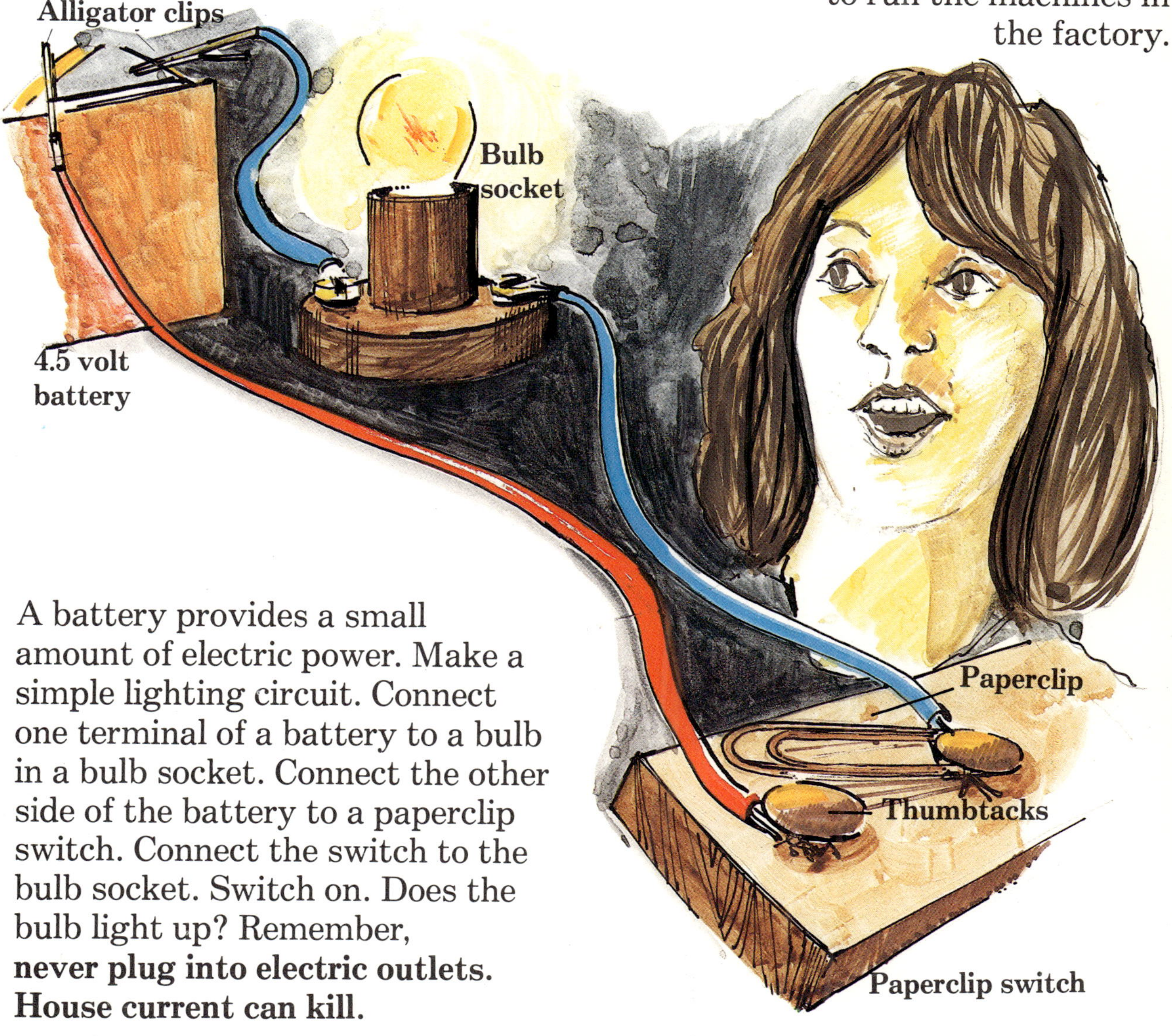

A battery provides a small amount of electric power. Make a simple lighting circuit. Connect one terminal of a battery to a bulb in a bulb socket. Connect the other side of the battery to a paperclip switch. Connect the switch to the bulb socket. Switch on. Does the bulb light up? Remember, **never plug into electric outlets. House current can kill.**

Make an electric motor

Many machines in factories are driven by electric motors. These motors use electricity and magnetism to make them work.
To make a simple motor, first find a small cork.
Cut two narrow grooves in opposite sides of the cork.
Push two small pins into one end, between the grooves of the cork.

Find some thin insulated wire and wind it, about thirty times, along the grooves of the cork.
Bare both ends of the wire and fix them to the small pins.

Find two long pins. Push them into the end of the cork to make a shaft for the motor.

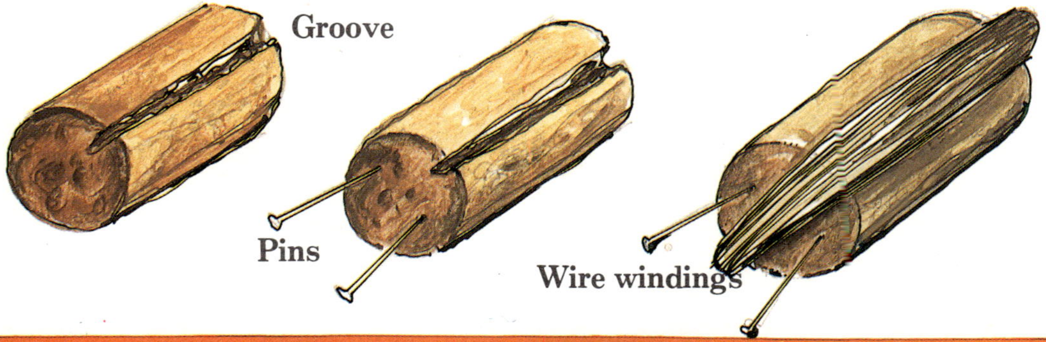

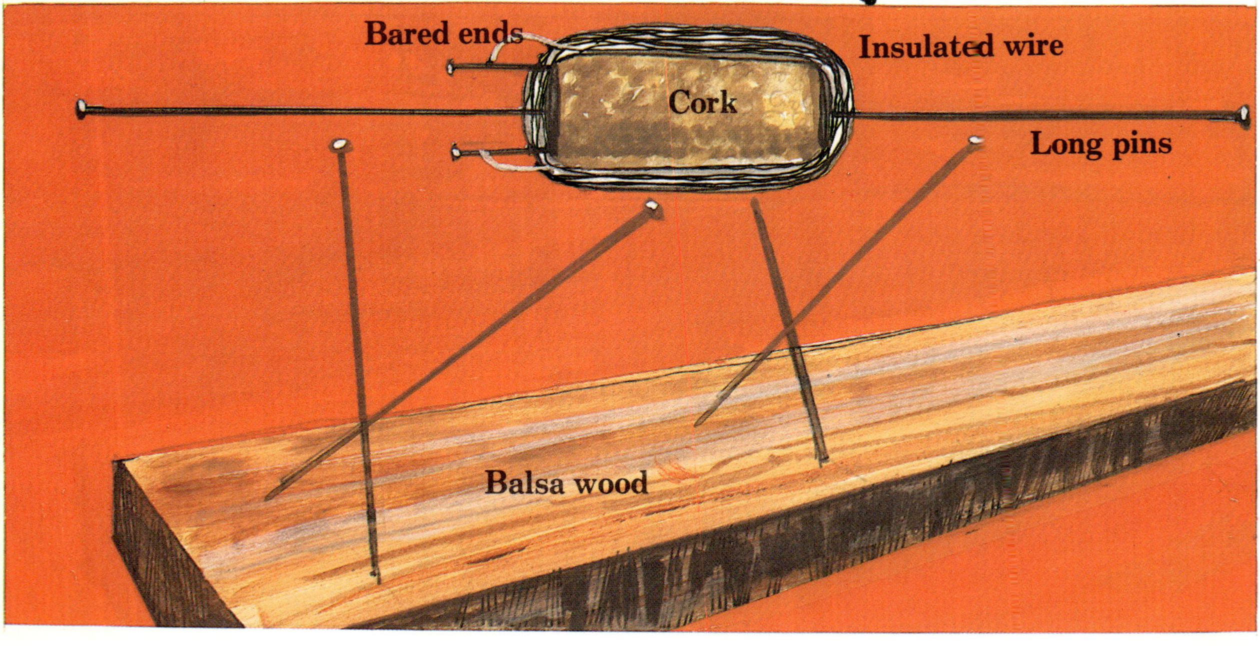

Use a piece of balsawood as a baseboard for your motor. Push four long pins into the balsawood to act as supports for the shaft.

Place the shaft on the pin supports. Cut two equal lengths of insulated wire, thicker than the one used on the cork. Cut the insulation from the ends of both wires. Use four thumbtacks to fix these wires to the baseboard. The ends of the wires nearest the cork should stand up vertically to brush the small pins in the cork. Connect alligator clips to the opposite ends of the wires.

Position the north pole of a strong bar magnet on a block of wood, close to the wire windings on the cork. Position the south pole of a bar magnet on the opposite side. Connect the alligator clips to a 4.5-volt battery. Give the shaft a slight turn to start your motor. Notice how the shaft revolves. It is this movement that is used to drive machinery.

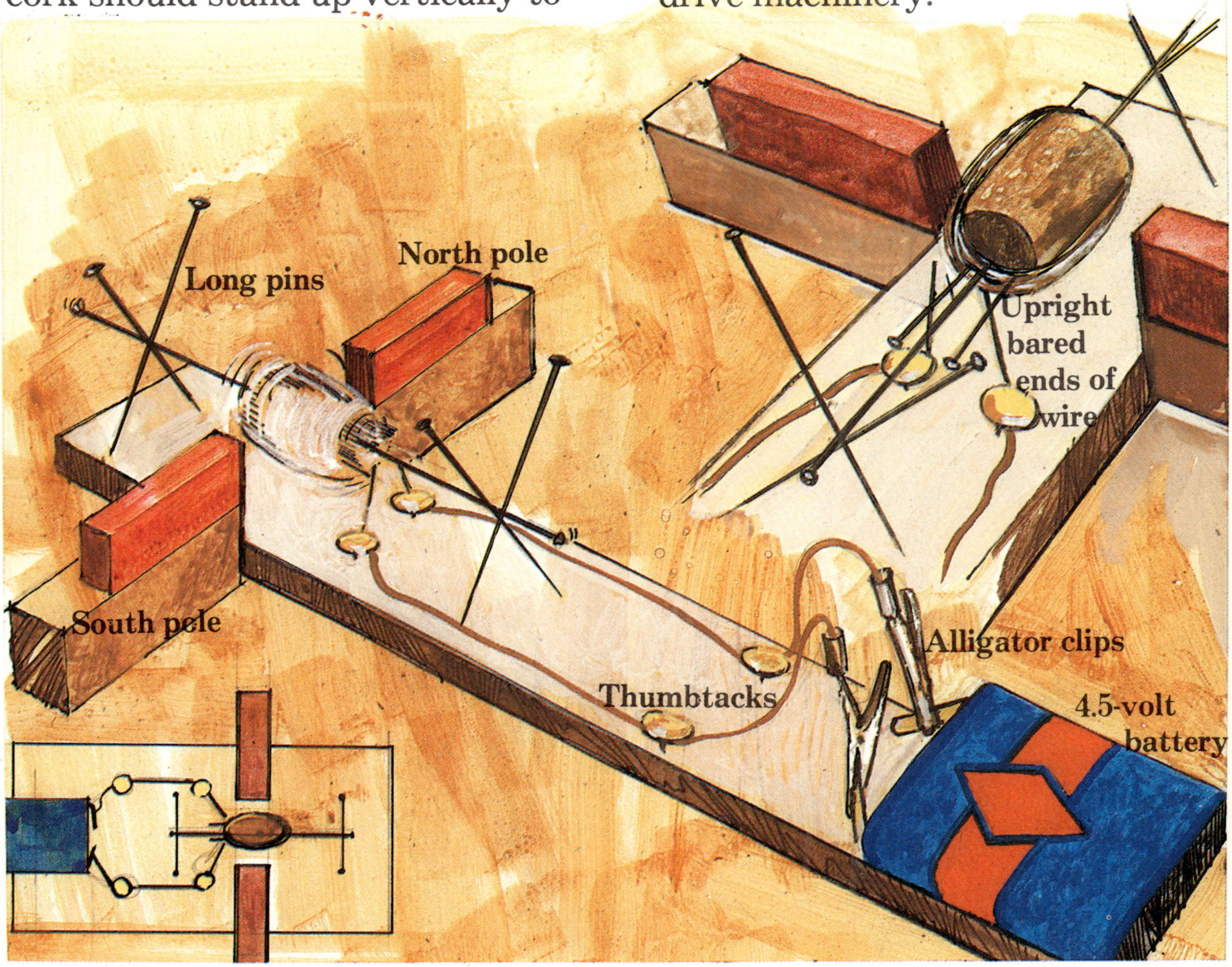

Compressed-air power

Find a bicycle pump. Pull out the handle of the pump. Air is drawn into the cylinder of the pump as you do this. Push in the handle and force the air out using the piston inside the cylinder. Pull out the handle. Place your hand over the end of the pump so that the air cannot escape. Push in the handle and squeeze the air in the cylinder. You are compressing the air. Take the pump to pieces and examine the parts.

A bicycle pump is a simple type of compressor. Larger compressors are used in factories. They are used to power drills, screwdrivers, paint sprayers and large machines.
Can you think of a way to compress air in a balloon to lift some weights in a yogurt container?

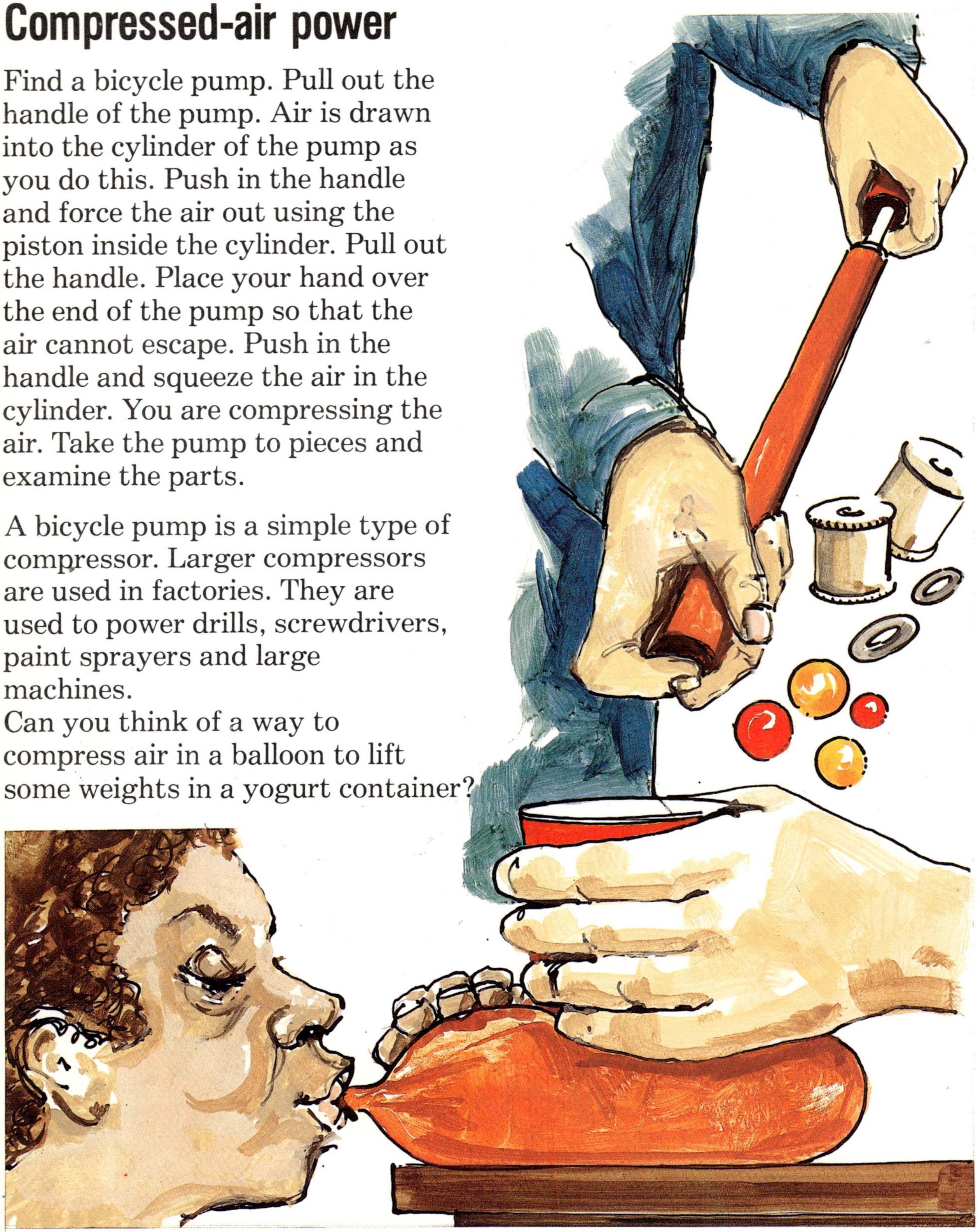

Chapter 3 Transferring power

Using levers

The lever is one of the oldest and simplest machines known. Find a can with a tight-fitting lid. Use a long screwdriver to lift the lid from the can. Push the end of the screwdriver under the lid. Rest part of the screwdriver on the edge of the can. Press down on the handle. Does the lid come off easily?

The lid is the load to be lifted. The screwdriver acts as a lever and pivots on the edge of the can. When you supply a force on the handle, this effort lifts the lid.

Look for ways we use levers at home, at school and in the garden. In the factory, levers have hundreds of uses. Tools such as wrenches and pliers are types of levers. Many machines use levers.

Using gears

A gear is a wheel with teeth around the edge. Collect three metal bottle caps. You can use these for your gear experiments. Make sure they are the same diameter. Find the exact center of each. Make a small hole in each center point. Use a small nail, through this hole, to fix one of the bottle caps to a piece of wood. Place the second cap next to the first so that the edges mesh together. Nail it in position. The bottle caps should revolve easily on their nail spindles. Now you can turn them like gear wheels.

Turn the first gear wheel. We can call this the driver gear. Notice how the gears mesh together. In which direction does the second, or driven gear, turn?
In a factory machine, the driver gear could be turned by an electric motor. The driven gear would then provide power in the opposite direction. Add a third gear alongside the other two. In which direction does it turn?

Gearbox of a car engine

Now build another gear system. Use two tin cans of different diameters. Find the center of the base of each can and make a small hole in each. Fix the larger can to a wooden baseboard. Use a nail through the center hole. Check that it revolves freely. Wrap some corrugated cardboard around the can and glue it in place.
Glue corrugated cardboard around the smaller can. Fix this can in position on the baseboard next to the large can.

Check that the corrugated cardboard "gears" mesh together.

Turn the smaller gear wheel. How much does the larger gear turn?

The shaft of an electric motor turns very quickly. It may be too fast to be useful in a factory. Can you think of how gears could reduce this speed?
Look at gear systems used in model toys.

If you have a construction kit, make a model using an electric motor and gear wheels.

Using belt drives

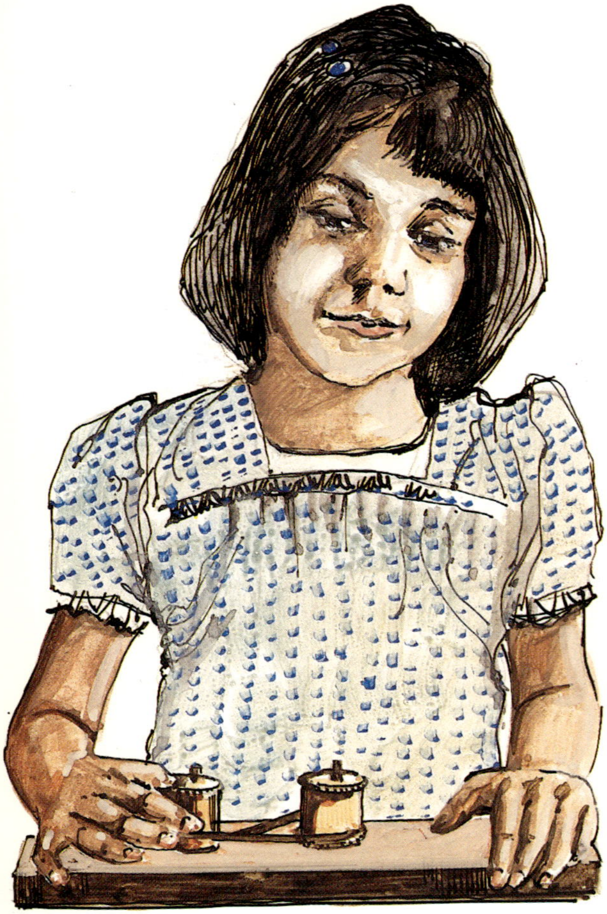

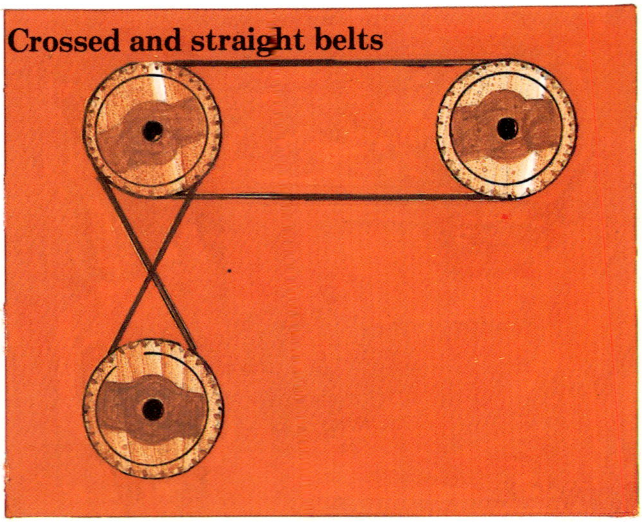

Find two wooden spools and fix them to a piece of wood with nails. Check that both spools can revolve easily. Hammer a small nail into the top of each spool. You can use these as handles to turn the spools. Slip a wide rubber band over both spools. Turn one of the spools in a clockwise direction.
What happens to the second spool? In which direction does it turn? Twist the rubber band to form a cross-over between the spools. Turn one of the spools.
What change do you notice? Build other belt-drive systems, using more spools.
Try using cans of different diameters, instead of spools.

Belt drives are often used to help electric motors turn machinery. Mount a small electric motor on a wooden baseboard. Make sure it is firmly fixed on the board.

Cut a length of dowel rod and push it through the center of an empty thread spool. It should be a tight fit.
Cut two rectangular pieces of balsawood.
Cut a V-shape from the top of each piece.
Nail these wooden pieces to the baseboard. Rest the dowel rod and spool in the V-shape.

Loop a rubber band around the spindle of the motor. Stretch it around the spool.
Connect the two leads from the motor to the terminals of a battery.
Is the electric motor turning the thread spool and the dowel rod shaft?
What happens if your drive belt is too loose?
Try using a crossed belt drive.

Using chain drives

Sometimes belt drives slip. This can be useful because it may prevent damage to machinery. When an engineer wants to avoid any slipping, a chain drive is used. Look at the chain drive on a bicycle.

Use the pedals to turn the large drive wheel. Notice that this wheel has a number of teeth on it. Engineers call this toothed wheel a sprocket. How many teeth are on this large sprocket? Count the number of teeth on the smaller, back-axle sprocket. Look at the chain. Notice the way it is made of a number of links. Notice how the links mesh with the teeth on the sprockets.

Chapter 4 Factory machines

A drop hammer

When metals need to be forced into a shape, some factories use a drop hammer. The piece of metal is heated until it is red hot. Then it is placed inside the drop hammer machine. A heavy hammer falls on the hot metal and shapes it. Many blows may be needed to get the correct shape. Cut some pieces of wood and make the framework for a drop hammer model. Find a large plastic bottle. Cut the top and bottom off the bottle to make a tube shape. Fasten this to the wooden framework.

Take an empty detergent bottle and fill it with sand. This is the hammer part of your machine.
Put a lump of modeling clay on the base of the wooden frame.
Drop your hammer, through the tube, onto the clay.
Can you change the shape of the clay?

Drilling

If you need to drill a hole at home, you may use a hand drill. Look at the part of the drill that actually makes the hole. It is called the drill bit. The cutting of the hole is done by the sharp cutting edges at the tip of the bit. These edges are made of very hard metals. They need to be kept sharp.

In factories, drilling holes is carried out by accurate drilling machines. The material to be drilled is placed on an adjustable worktable. This holds the material still while a revolving drill head descends onto it. Sometimes one machine may have many drill heads. This means that more than one hole can be drilled at the same time.

Drill bit

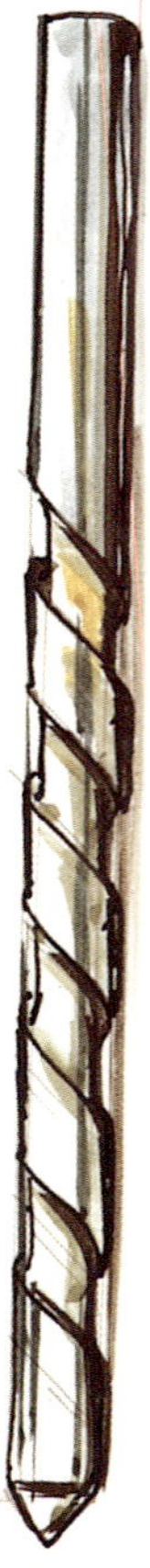

Cutting edge

Lathes

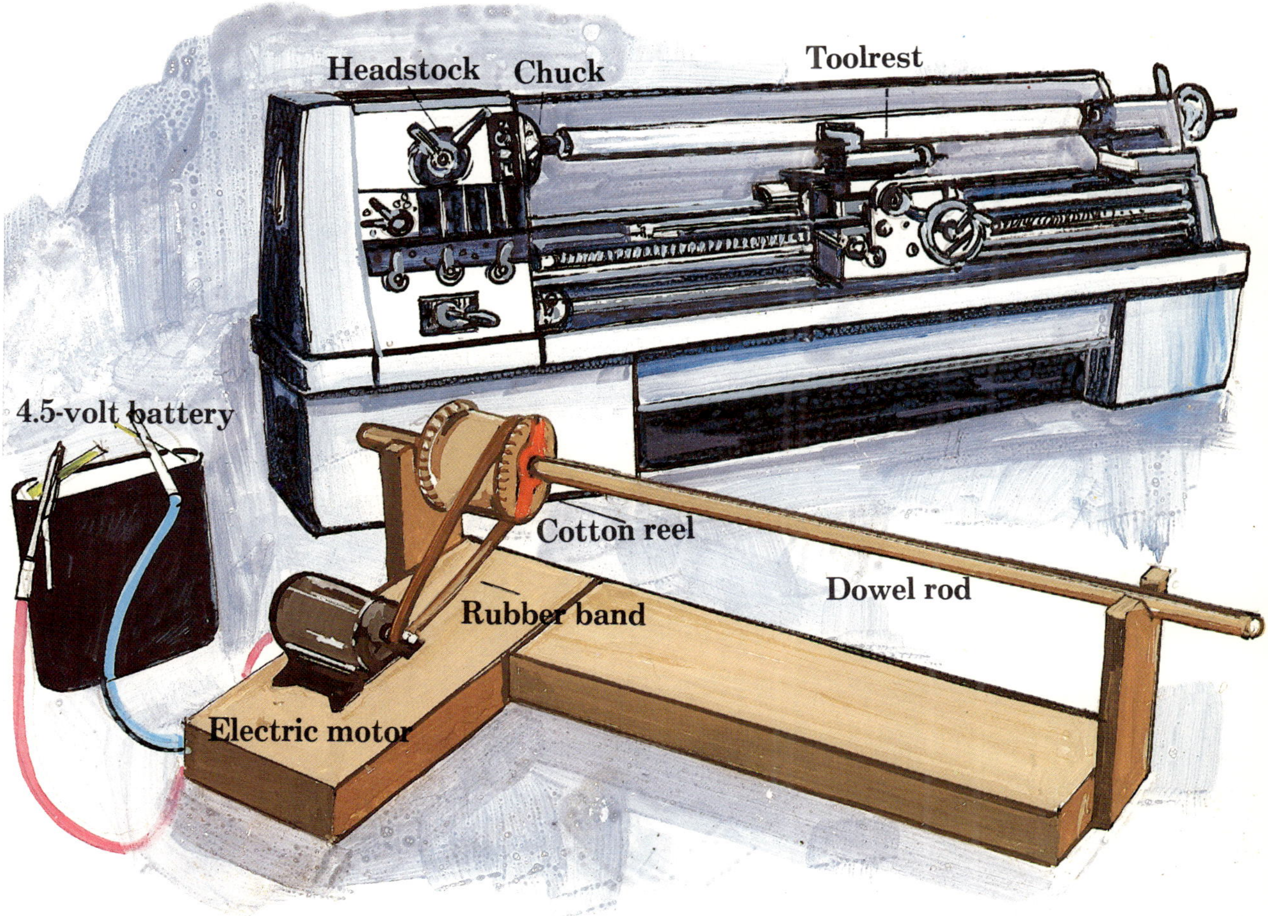

The lathe is one of the most important machines used in factories.

The headstock of the lathe contains an electric motor and a gearbox. This turns the chuck at different speeds. The material to be worked is placed in the jaws of this chuck. As the material rotates in the chuck, it can be shaped by the worker using the toolrest.

Make a simple lathe. Push a length of dowel rod through a spool. Make sure it is a tight fit. Mount this rod on two pieces of wood. You will need to cut V-shapes out. Connect the spool to a small electric motor by using a belt drive (see page 17).

Connect the motor to a battery. Does the dowel rod rotate in the spool chuck?

Robots

Robots are being used in more and more factories. They can do work that humans find boring and unpleasant. Often these jobs may be very simple. In car factories robots are used for painting and joining parts of car bodies together. They work quickly and accurately.

These robots need to be given instructions before they can work. The instructions are stored in the memory of the robot's computer. Some modern factory robots can obey over twenty different sets of instructions. The robot carries out the work, following the instructions it has been given. Many robots use electric motors or compressed air to move their parts.

Do you have a computer at home or in school?

How do you give instructions to your computer?

How do you give it a new set of instructions?

Make a simple electromagnet robot arm of your own.
Find a long cardboard tube.
Make a small hole near one end of the tube.
Fit a short length of dowel through this hole.
Push an empty spool on the other end of the dowel. Press a nail through the end of the tube.
Can you swivel your robot arm on its dowel support?

Take a long length of covered wire and wind part of the wire around the nail. Make about fifty turns around the nail.
Push both ends of the wire through the tube. Connect these ends to a battery and a paperclip switch (see page 9).

Switch on.
Does your electromagnet pick up a paperclip?

Use your robot arm to transfer some paperclips from one yogurt container to another.

Chapter 5 Moving goods

Cranes

Very heavy loads are moved around factories by cranes. These cranes move on tracks above the working area. Steel girders form a movable bridge across the factory. The crane hoist is carried in a trolley that can move across the steel bridge. This type of crane can reach any place on the factory floor that is between the tracks on the factory walls.

You can make a bridge crane. Find a length of wood to use as the bridge across the working area. Find a smaller piece of wood to act as the trolley. This piece should be slightly longer than the width of the bridge.
Glue two strips of wood along the edges of the trolley piece.

Fit your trolley on top of the bridge.
Does it slide easily along the bridge?

Mount a small electric motor, a battery, a belt drive and a spool on the trolley (see page 17). Push a thumbtack into the spool. Fasten a length of thread to this tack. Make a hook from a paperclip and fasten this to the thread.

Position your bridge crane across two chairs. Move the trolley from side to side. Connect the motor terminals to the battery. Does the motor turn the spool? Does the thread wind around the spool? Connect the battery terminals the opposite way around? What happens?

Fork-lifts

Many modern factories use trucks with "fork-lifts" to lift and stack goods. The goods are placed on wooden platforms called "pallets." The front of the truck has two metal forks. These slide under the pallet carrying the goods. The forks, pallet and goods are then raised by a hydraulic jack. Many trucks use batteries to power electric motors. These motors move the trucks backward and forward in the factory.

Try to watch a fork-lift truck at work.
Notice the metal frame around the driver. What use is this frame?
How does the driver steer the truck?
Which wheels are used for steering?

Make a simple hydraulic lifting system. Take a large plastic bottle and cut it in half. Keep the bottom half of the bottle. Cut a length of rubber tubing. Connect it to the end of a plastic syringe.

Make a small hole near the bottom of the half-bottle container. Fit the rubber tubing into the hole. Make sure it is a tight fit.

Pour water into the plastic container so that it enters the tubing and the syringe. Take an empty yogurt container and place it on the water. Watch the yogurt container as you press in the plunger of the syringe. Put a spool in the yogurt container. Can you still lift it?

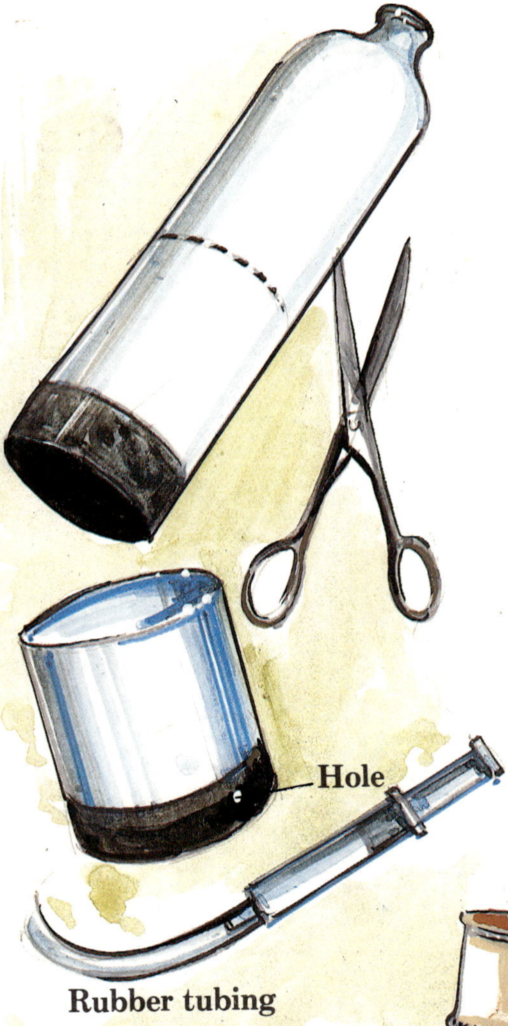

Conveyors

Conveyors are used to move raw materials and finished products in a factory.
The simplest type of conveyor makes use of the force of gravity to move goods.
Find two plastic detergent bottles.
Cut off both ends from each bottle.
Take both tube sections and cut them in half.
Use tape to join the four pieces of plastic together to make a slide.

Place some books on the floor.
Rest one end of your slide on the books, so that it slopes.

Test your conveyor using pencils, erasers, spools and other small items.
Do they travel down the slide?

A gravity conveyor can easily be moved from place to place in a factory.

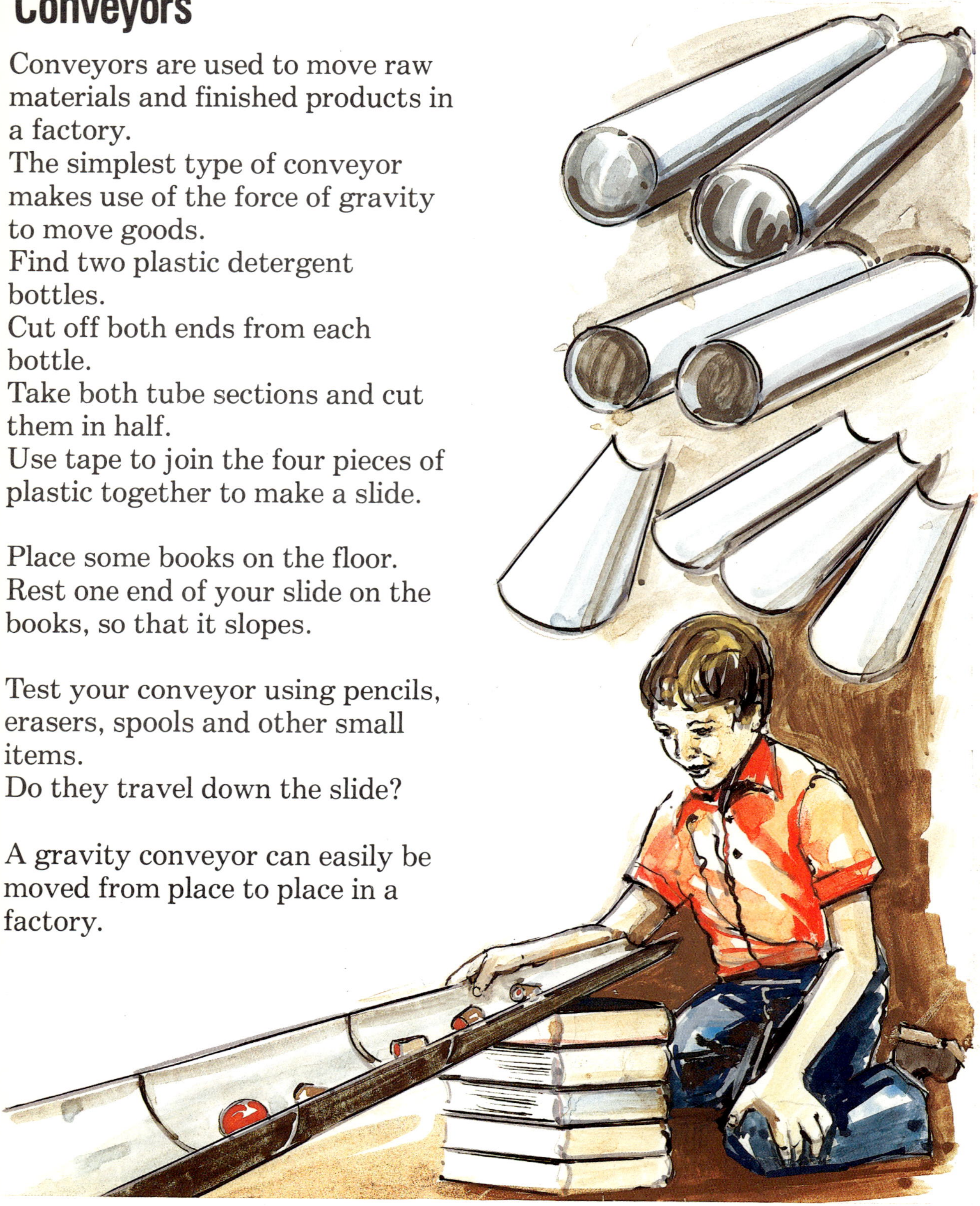

When goods need to be moved uphill or along the level, a belt conveyor may be used. Use your electric-drive belt model (see page 17) to make a belt conveyor.

Use wood, an empty thread spool and a dowel rod to make a second supported roller. Let the spool rollers revolve on the dowel rod shafts. Find a long, wide rubber band. Position it over both rollers. Connect the electric motor to the battery. Notice how the motor turns the first roller. This roller then turns the conveyor belt.

Many factories use conveyors to help assemble their products. Workers are positioned at different points along the conveyor. Each worker carries out a different task on the product. In some factories these assembly lines use overhead conveyors.

Chapter 6

Factory problems

Friction

Find a hand-drill and a small piece of dowel rod.
Fit the dowel rod into the hand-drill.
Use this dowel "bit" to drill into a block of wood.
Feel the end of the dowel rod.
Is it warm?
Now rub your hands together.
Do they begin to feel hot?

When two surfaces move over one another, friction occurs. This force tries to slow down the movement and produces heat. Friction causes great problems for the engineer in the factory.
Friction from the moving parts of the machines will produce heat and cause wear. Engineers use lubricants, such as oil and grease, to try to reduce friction.

Keeping cool

Inside some factories it may become very hot.
Can you think of reasons for this?
The engineer must try to lower the temperature, so that the people in the factory are comfortable.
One way is to use cooling fans.

Make a cooling fan of your own. Fit a plastic propeller to the spindle of a small electric motor.

Use balsawood for a tight fit if the spindle is smaller than the hole in the propeller.
Mount the motor and propeller on top of a piece of wood. Fix this piece of wood to a baseboard.
Connect two leads from the motor to the terminals of a battery.
Can you feel a draft from your fan?
Connect a paperclip switch into the cooling fan circuit (see page 9).

Safety

Safe working conditions are important to the people who work in factories. There are laws to help protect these workers. Engineers need to know the laws and to see that they are obeyed. Moving machinery can be very dangerous. It is often protected by wire netting.

Some factories are noisy. Workers may need to wear ear protectors.

Others must wear goggles to protect their eyes.

If there is a lot of dust, they will wear special face masks.

In many modern factories people work in clean and comfortable surroundings.

Engineers must also think of the people outside the factory. The factory must not pollute the air around it. Look for smoke coming from factory chimneys.
What will happen if waste materials and liquids are emptied into rivers and onto the land?

Imam Ali (A)
2

Heavenly Children

The Brave Defender

Kisa Kids Publications

Dedication

This book is dedicated to the beloved Imām of our time (AJ). May Allāh (swt) hasten his reappearance and help us to become his true companions.

Acknowledgements

Prophet Muḥammad (s): The pen of a writer is mightier than the blood of a martyr.

True reward lies with Allāh, but we would like to sincerely thank the efforts of Shaykh Salim Yusufali, Brother Aliakbar Shaheidari, Sisters Sabika Mithani, Fatemah Mithani, Amna Hussain, Asieh Zarghami, Zahra Sabur, Sajeda Merchant, Kisae Nazar, Sarah Assaf, Nadia Dossani, Fathema Abidi, Fatemeh Eslami, Fatima Hussain, Fatemah Meghji, Sukaena Kalyan, and Zehra Abbas. We would especially like to thank Nainava Publications for their contributions. May Allāh bless them in this world and the next.

Preface

Prophet Muḥammad (s): Nurture and raise your children in the best way. Raise them with the love of the Prophets and the Ahlul Bayt (a).

Literature is an influential form of media that often shapes the thoughts and views of an entire generation. Therefore, in order to establish an Islāmic foundation for the future generations, there is a dire need for compelling Islamic literature. Over the past several years, this need has become increasingly prevalent throughout Islāmic centers and schools everywhere. Due to the growing dissonance between parents, children, society, and the teachings of Islām and the Ahlul Bayt (a), this need has become even more pressing. Al-Kisa Foundation, along with its subsidiary, Kisa Kids Publications, was conceived in an effort to help bridge this gap with the guidance of ʿulamah and the help of educators. We would like to make this a communal effort and platform. Therefore, we sincerely welcome constructive feedback and help in any capacity.

The goal of the *Heavenly Children* series is to foster the love of Ahlul Bayt (a) in children and to help them establish the 14 Maʿṣūmīn as their role models. We hope that you and your children enjoy these books and use them as a means to achieve this goal, inshāʾAllāh.

We pray to Allāh to give us the strength and *tawfīq* to perform our duties and responsibilities.

With Duʾās,
Nabi R. Mir (Abidi)

Disclaimer: Religious texts have ***not*** been translated verbatim so as to meet the developmental and comprehension needs of children.

Kisa Kids Publications
4415 Fortran Court
San Jose, CA 95134
(260) KISA-KID [547-2543]

Ahmad's Mama and Baba had just shared some exciting news with him: they would be going on a trip to visit all the fourteen Ma'soomeen. Mama had just told him about Prophet Muhammad (s) as a child.

"Mama, will we also be able to visit Imam Ali (a)?" asked Ahmad.

Mama beamed. "Yes, of course! We will visit him in Najaf, where he is buried, InshaAllah!" she replied.

"Alright!" exclaimed Ahmad excitedly. "Baba, I remember Imam Ali (a) was Prophet Muhammad's (s) biggest supporter, but can you please remind me more about him?"

His father answered, "Sure, my son, it would be my honor! Imam Ali (a) is our first Imam, remember? He was the cousin of Prophet Muhammad (s), and he was also married to Prophet Muhammad's daughter. He was born on Friday, the 13th of Rajab, and he was born in a place where nobody else has ever been born!"

"I know! The Holy Ka'bah!" Ahmad was so excited to know this exceptional fact.

"That's right! MashaAllah. Allah allowed the wall of the Ka'bah to crack open so that his mother, Fatimah bint* Asad, could enter. She was Allah's special guest for three days. She came out with Imam Ali (a) in her arms, and presented him to his father, Abu Talib."

Ahmad thought for a minute. "So, how come he's not buried near the Ka'bah?"

Mama replied, "Well, Imam Ali (a) had moved to a place called Kufa to lead the Muslim ummah. Sadly, he was struck with a sword while he was in sajdah during Fajr prayers in Masjid al-Kufa on the 19th of Ramadhan and became a shaheed** three days later on the 21st of Ramadhan. That's why he's buried nearby in Najaf."

"Oh," said Ahmad softly. Suddenly a question came to his mind. "I know Imam Ali (a) was Prophet Muhammad's supporter, but can you tell me how?"

"Well," Baba said "firstly, he was very, very brave and always stood up for the truth. You see, Imam Ali (a) was basically raised by Prophet Muhammad (s) in his home. So, he learned the best akhlaq. Like the Prophet (s), he was always fair, honest, caring, content, and polite. He treated everyone in the best way. That is why Allah chose him to be the successor of Prophet Muhammad (s). Let me tell you a story of how Imam Ali (a) was brave, even as a young boy."

*Daughter of

**Shaheed: someone who dies in the path of Allah

The sun was setting over the city of Mecca as the Noble Prophet (s) entered his home. His body was covered with cuts and bruises, and blood was dripping from his forehead. The Prophet (s) went out to the backyard and used a clay jug of water to wash the bloodied wounds on his face.

As soon as Imam Ali (a) heard the Prophet (s) enter the home, he eagerly ran outside to greet him. His happiness soon faded when he saw what had happened to the Prophet (s)! Imam Ali (a) quickly ran back inside to bring a towel for the Prophet (s) to dry off with.

As the Prophet (s) wiped his face, he could see that the young Imam was alarmed and upset to see him in this condition. Prophet Muhammad (s) gave Imam Ali (a) a reassuring smile. Still, Imam Ali (a) let out a sad sigh. He knew exactly what had happened; the enemies of Islam had hurt his beloved Prophet (s) once again.

The Prophet’s enemies had instructed their children to bother the Prophet (s) as much as they could. Every time Prophet Muhammad (s) would step outside, the children would gather in groups, throw stones, and make fun of him. They did not understand who the Prophet (s) was or what his mission was. They just followed what the misguided people told them to do.

Ever since the Prophet's uncle, Abu Talib, had passed away, no one had been able to protect the Prophet (s). It had been Abu Talib who had kept the Prophet (s) safe from the enemies of Allah. Abu Talib was a respected elder in Mecca, so while he was alive, nobody dared to hurt the Prophet (s). Now, Abu Talib was not there.

Imam Ali (a) wished that he could be just like his father, Abu Talib, and keep the Prophet (s) safe from harm.

Seeing the Prophet (s) in so much pain made Imam Ali's heart ache. How could the people do this to the best of Allah's creation and kindest of human beings? As he glanced at the Prophet's face, Imam Ali (a) could see the sadness in his eyes.

He had to do something. Slowly, he asked, "Dear Rasulullah (s), can you please take me with you the next time you decide to leave the house?"

The Prophet (s) smiled and lovingly placed his hands on Imam Ali's shoulders. He nodded in agreement. "Okay, my beloved."

Prophet Muhammad (s) was very proud of Imam Ali (a). Even though he was just a young boy, he was as brave as a warrior!

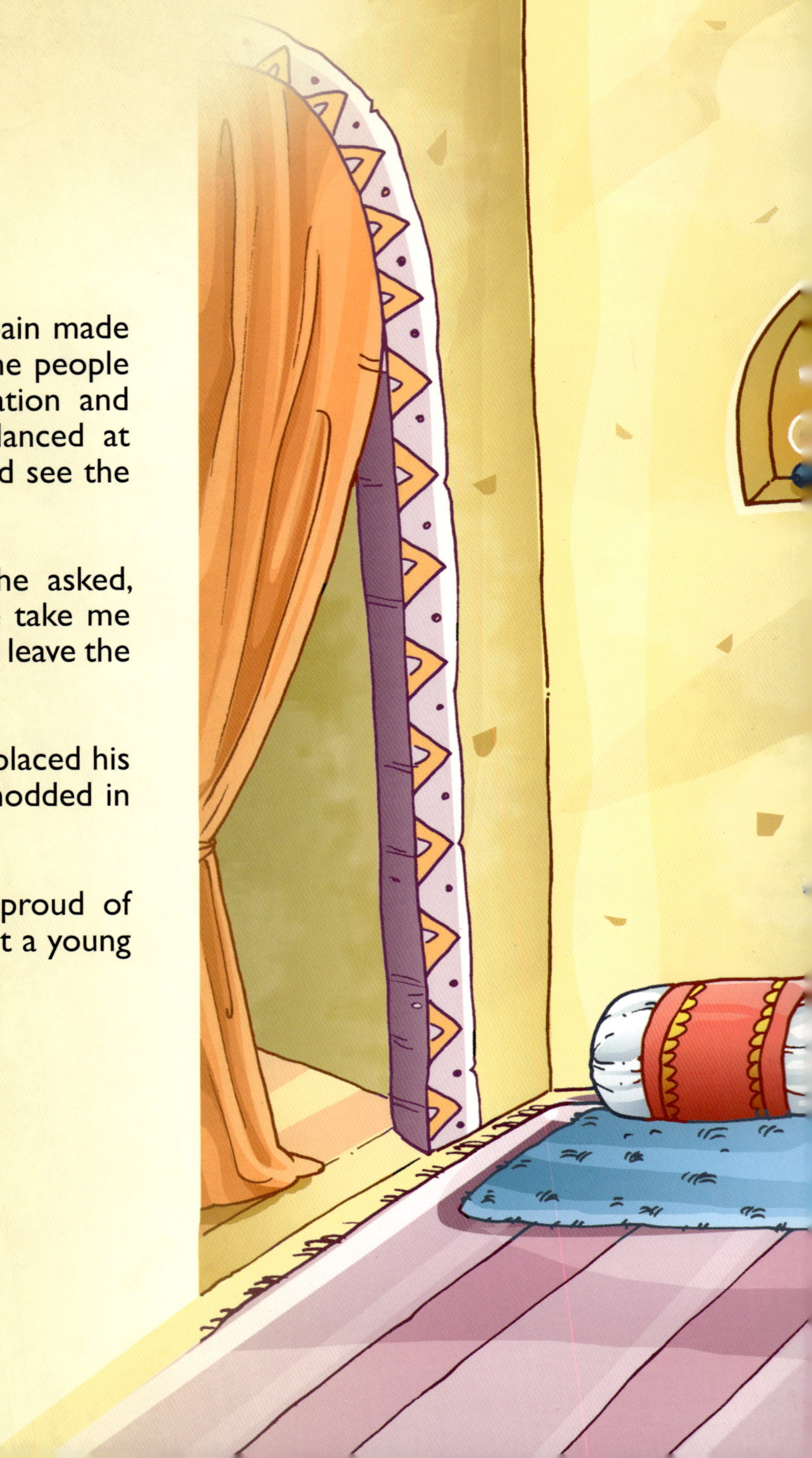

As night fell, Imam Ali (a) lay in bed, but he could not sleep. He sat up. He kept thinking about the pain and sadness he had seen in the Prophet's eyes that day. How could the enemies torture his beloved Prophet (s)? Everyone knew how kind and trustworthy the Prophet (s) was!

In that very moment, he made a decision: he would never let anyone harm the Prophet (s) ever again! Only after making this promise was he able to fall peacefully asleep.

The next morning, as the Noble Prophet (s) got ready to go out for the day, he found Imam Ali (a) waiting for him, dressed and ready to go. The Prophet (s) gazed at Imam Ali (a) in awe. This young boy was only ten years old, but he looked much stronger and bigger than his age and his eyes were full of courage. Looking at Imam Ali (a) gave Prophet Muhammad (s) an extra boost of strength.

As the Prophet (s) left his home, Imam Ali (a) walked right next to him, keeping a watchful eye. It was not long before they came across a group of kids playing in the streets.

One of the boys saw the Prophet (s) and teased, "Oh, look, it's Muhammad!" The other kids began to chime in, "Crazy, crazy, crazy, crazy!"

The Prophet (s) and Imam Ali (a) kept walking, but as they got closer, the children started yelling louder and picking up stones.

One boy gripped a stone in his fingers and lifted his hand back, poised to throw. But just as he pitched forward to throw the rock, a firm hand grasped his wrist, twisting his arm and forcing him to unclench his fingers. The stone fell to the ground. Shocked, the boy looked up and saw that it was the young Imam Ali (a). His face quivered with fear.

As the boy stumbled back, Imam Ali (a) bolted towards the rest of the children, who had not noticed this young challenger in their midst. They were still yelling and gathering stones to pelt at the Noble Prophet (s).

Imam Ali (a) tackled them all! He was able to wrestle the stones away from every kid. Soon, they were all on the ground, sobbing and moaning.

Imam Ali (a) boldly asked them, "Anybody else want to throw rocks?!"

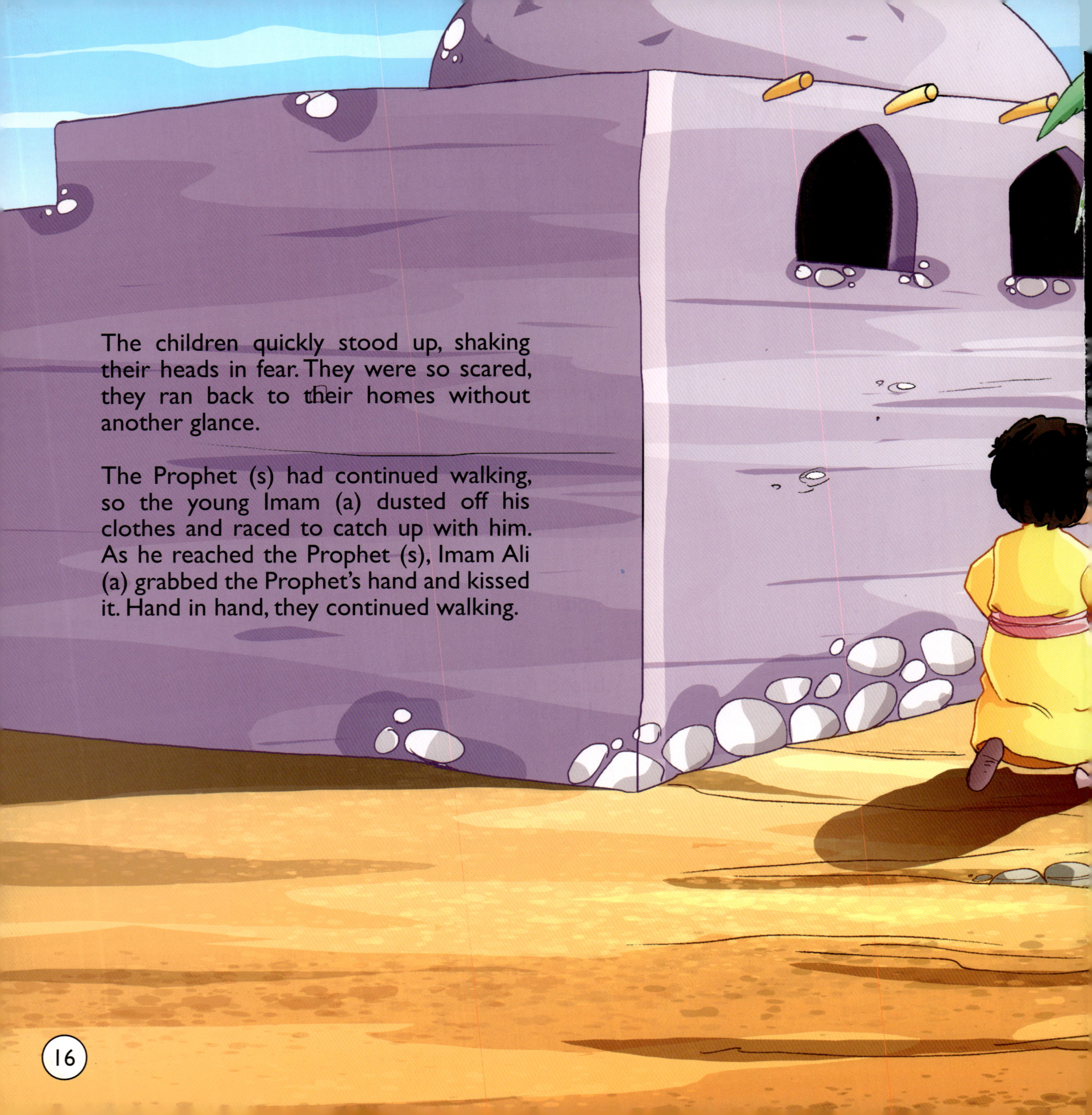

The children quickly stood up, shaking their heads in fear. They were so scared, they ran back to their homes without another glance.

The Prophet (s) had continued walking, so the young Imam (a) dusted off his clothes and raced to catch up with him. As he reached the Prophet (s), Imam Ali (a) grabbed the Prophet's hand and kissed it. Hand in hand, they continued walking.

From that day forward, Imam Ali (a) tried his best never to leave the Prophet's side.

Every day, the children of Mecca would see Imam Ali (a) walking alongside the Prophet (s). Yet, each and every day, all they could do was look at them silently because they no longer dared to even shout at the Prophet of Islam (s).

A'yān ash-Shī'ah, Vol. 1, P. 372